"Audits and Corrective Action: Essential Tools for Business Success"

RODRIGO PALMA MENA

Copyright © 2024 Rodrigo Palma Mena

All rights reserved.

DEDICATION

Dear Family,

I want to dedicate this book to all of you, who have been my support and strength throughout this time. Without your love and support, it would not have been possible to achieve this achievement.

Thank you for your patience, understanding and constant motivation. Each of you has been a pivotal pillar in my life and I am grateful to have you in my family.

I hope this achievement inspires each of you to follow your dreams and goals with dedication and effort. Always remember that the love and support of family is the key to success.

With all my love and gratitude,

Rodrigo R. Palma Mena

Content

1 INTRODUCTION

2 INTRODUCTION TO MANAGEMENT SYSTEM AUDITS

    OBJECTIVES AND PRINCIPLES OF AUDITS

    IMPORTANCE OF INTERNAL AUDITS FOR QUALITY MANAGEMENT

3 NORMATIVE AND CONCEPTUAL FRAMEWORK

    LEGAL AND REGULATORY REQUIREMENTS FOR INTERNAL AUDITS

    KEY CONCEPTS IN AUDITS AND QUALITY MANAGEMENT

4 INTERNAL AUDIT PLANNING

    REQUIREMENTS AND CRITERIA FOR AUDIT PLANNING

    DEVELOPMENT OF AN INTERNAL AUDIT PROGRAM

5 PREPARATION AND EXECUTION OF AUDITS

    SELECTION AND TRAINING OF INTERNAL AUDITORS

    AUDIT EXECUTION PROCESS: INITIATION, PREPARATION AND IMPLEMENTATION

6 EVIDENCE COLLECTION AND EVALUATION

    TECHNIQUES AND METHODS FOR COLLECTING EVIDENCE DURING THE AUDIT

    OBJECTIVE EVALUATION OF EVIDENCE TO IDENTIFY NON-CONFORMITIES

7 IDENTIFICATION AND CLASSIFICATION OF NON-CONFORMITIES

    DEFINITION AND TYPE OF NON-CONFORMITIES

CRITERIA FOR CLASSIFYING AND PRIORITIZING THE NON-CONFORMITIES FOUND ....

8 CORRECTIVE AND IMPROVEMENT ACTIONS ....

DEVELOPMENT OF CORRECTIVE AND IMPROVEMENT ACTIONS ....

EFFECTIVE IMPLEMENTATION OF ACTIONS TO RESOLVE NON-CONFORMITIES ....

9 MANAGEMENT AND MONITORING OF NON-CONFORMITIES ....

NON-CONFORMANCE MANAGEMENT PROCESS ....

FOLLOW-UP AND CLOSURE OF NON-CONFORMITIES TO ENSURE THE EFFECTIVENESS OF ACTIONS ....

10 FOLLOW-UP AUDITS AND SYSTEM REVIEW...................

IMPORTANCE OF FOLLOW-UP AUDITS IN CONTINUOUS IMPROVEMENT..

REVIEW OF THE MANAGEMENT SYSTEM BASED ON AUDIT FINDINGS ....

11 CONTINUOUS IMPROVEMENT AND ORGANIZATIONAL LEARNING..

USING LESSONS LEARNED FROM AUDITS TO IMPROVE PROCESSES AND SYSTEMS...............................

INTEGRATION OF CONTINUOUS IMPROVEMENT INTO ORGANIZATIONAL CULTURE ....

12 CASE STUDIES AND EXAMPLES..

DETAILED ANALYSIS OF REAL CASES OF INTERNAL AUDITS AND TREATMENT OF NON-CONFORMITIES ACCORDING TO ISO19011:2018 ....

13 CONCLUSION ....

14 BIBLIOGRAPHY...

THANKS

I, Rodrigo Palma Mena, would like to take this opportunity to express my sincere gratitude to all the people who have given me their support and help in the realization of this project.

First of all, I want to thank you
to Renato and Sebastián Palma for their constant support and encouragement in moments of doubt and fatigue during the preparation of this book. Thanks to them, I have been able to overcome obstacles and move forward.

Finally, I thank my family and friends for their understanding and encouragement throughout, and for their confidence in my ability to carry out this project.

To all of them, my sincere thanks. This book would not have been possible without your unconditional collaboration and support.

1 INTRODUCTION

Rodrigo Palma Mena is a consultant born in Santiago de Chile in 1977. He is known for advising companies that choose to implement quality standards in different areas, for teaching regulatory courses requested by accreditation bodies in Chile, as well as training and preparing quality managers and competent auditors. In addition to his consulting career, he has worked as a chemist and quality engineer in different companies in different areas such as pharmaceuticals, raw materials, pesticides, herbicides, clinics and mining. He has more than 8 years of experience related to leading and training people, always focused on continuous improvement.

In today's dynamic business environment, quality and operational excellence are crucial elements to the success of any organization. The ability to effectively identify, address and resolve non-conformities in management systems has become a determining factor in maintaining high quality standards and meeting customer and stakeholder expectations.

This book on internal audits and treatment of non-conformities is designed as a practical and comprehensive guide, based on the principles and guidelines established by the ISO 19011:2018 standard, which provides guidelines for the audit of management systems. Our goal is to provide readers, whether they are quality professionals, internal auditors, management system managers or managers, with the tools and knowledge necessary to carry out effective audits and efficiently manage detected non-conformities.

Internal audits are an essential component of any effective management system. They make it possible to assess the conformity of systems, identify areas for improvement, and provide objective feedback for the strengthening of the organization. Beyond simple compliance assessment, internal audits foster a culture of continuous improvement by providing opportunities to correct deviations and optimize processes.

ISO 19011:2018, guidelines for the audit of management systems, sets out the principles and best practices for conducting effective and efficient internal audits. This standard is applicable to quality, environmental, safety, and other management systems, providing a common framework for planning, execution, and monitoring of audits.

The main purpose of this book is to equip readers with the knowledge and skills necessary to:

- Plan, prepare and execute robust and results-oriented internal audits.
- Identify, classify, and appropriately address non-conformances detected during audits.
- Develop and implement effective corrective and preventive actions to improve management systems.
- Comprehensively manage non-conformities to ensure continuous improvement and regulatory compliance.

This book is organized into a series of chapters covering various aspects related to internal audits and the treatment of non-conformities. We start with the fundamental concepts and principles of auditing, moving towards the planning and execution of audits, and finally explore in detail the management of non-conformities and continuous improvement. Each chapter is designed to offer a balanced mix of theory and practice, supported by real-world examples and helpful tips.

This book is aimed at a broad audience, including, but not limited to:

- Quality professionals and internal auditors responsible for management systems.
- Directors and managers committed to continuous improvement and operational excellence.
- Students and academics interested in understanding best practices in internal audits and non-conformance management.
- Anyone involved in the implementation and maintenance of management systems based on international standards such as ISO 9001, ISO 14001, ISO 45001, ISO17025; ISO17020, among others.

Each chapter of this book is designed to provide a practical and accessible approach for readers. From a clear introduction to key concepts to detailed examples and practical recommendations, this book will serve as a valuable reference and learning tool for those looking to improve their skills in internal audits and non-conformance management.

Without further ado, let's start discovering how internal audits and effective handling of non-conformities can boost your organization's quality, efficiency, and competitiveness. We will explore best practices and strategies to achieve operational excellence in an increasingly demanding and changing world.

2 INTRODUCTION TO MANAGEMENT SYSTEM AUDITS

OBJECTIVES AND PRINCIPLES OF AUDITS

Management system audits are a critical component of ensuring that organizations operate effectively, efficiently, and in compliance with established standards and requirements. These audits are systematic and independent processes that make it possible to evaluate the effectiveness and adequacy of the management systems implemented.

The main purpose of management system audits is to identify opportunities for improvement and ensure compliance with the organization's objectives and goals. By verifying compliance with internal standards, legal standards, and regulatory requirements, audits provide a critical and objective assessment of organizational performance.

Management system audits can cover a variety of areas, including quality systems (ISO 9001), environment (ISO 14001), occupational health and safety (ISO 45001), information security (ISO 27001), among others. Each management system requires periodic evaluation to ensure its effectiveness and the achievement of the organization's strategic objectives.

During an audit, processes, procedures, records, and operational practices are reviewed and evaluated to identify areas for improvement, potential risks, and opportunities to optimize overall performance. Auditors, who may be internal or external to the organization, must possess the technical competence and impartiality necessary to carry out an objective and fair assessment.

In today's business landscape, where quality, operational efficiency, and customer satisfaction are key imperatives, management system audits play a critical role in the evaluation and continuous improvement of organizations. Audits not only seek to verify compliance with specific requirements, but also aim to provide an objective and evidence-based assessment of the performance of management systems.

Objectives of the Audits

Assess Conformity:

The primary objective of an audit is to determine whether the audited management system complies with the requirements established by applicable regulations, internal and external standards, organizational policies, and other criteria.

Conformity assessment involves verifying the effective implementation of the processes and controls necessary to meet the established requirements.

Identify Areas for Improvement:

Audits seek to identify opportunities for improvement in management systems by critically reviewing current practices.

The identification of areas for improvement can cover aspects such as operational efficiency, effectiveness of internal controls, risk management, customer satisfaction, among others.

Provide Objective Feedback:

Audits provide an unbiased, evidence-based assessment of management system performance.

Objective feedback obtained during an audit allows informed decisions to be made to improve processes and systems.

Verify System Effectiveness:

Audits aim to verify the effectiveness of the management system in achieving the planned and expected results.

This verification includes making sure that the objectives of the system are being achieved effectively and that the expectations of stakeholders are being met.

Promoting Transparency and Trust:

Audits foster transparency by ensuring that organizational processes and practices are reviewed systematically and objectively.

Transparency promotes trust among stakeholders, including customers, employees, suppliers, and accreditation bodies.

Principles of Audits

Integrity:

Auditors must act with integrity, impartiality and professional ethics at all stages of the audit.

Integrity ensures that audit findings are objective and free from undue influence.

Evidence-Based Approach:

Audit conclusions should be based on verifiable and objective evidence obtained during the audit process.

The evidence-based approach ensures the reliability and credibility of the audit's findings.

Confidentiality:

Information collected during an audit should be treated confidentially to protect the interests of the audited organization.

Confidentiality is essential to maintain the integrity and objectivity of the audit process.

Risk Management Approach:

Auditors should consider and assess the risks associated with the audited management system.

The risk management approach makes it possible to identify critical areas that require attention and corrective actions.

Life Cycle Approach:

Audits should cover the entire lifecycle of the management system, from planning to implementation and follow-up of corrective actions.

This comprehensive approach ensures that all stages of the management system are properly assessed.

Continuous Improvement Approach:

Audits should promote continuous improvement by identifying opportunities for improvement and recommending effective corrective actions.

Continuous improvement is essential to optimize the performance and effectiveness of the management system.

Systems Approach:

Audits should assess the management system as a whole, considering the interactions between processes and functions within the organization.

The systems approach makes it possible to identify synergies and areas for improvement that can benefit the entire organization.

These objectives and principles guide the conduct of effective audits and contribute to strengthening management systems to achieve optimal levels of performance and compliance. The application of these principles promotes confidence in audit results and their ability to drive continuous improvement in organizations.

IMPORTANCE OF INTERNAL AUDITS FOR QUALITY MANAGEMENT

Internal audits are not only a critical tool for assessing compliance with established requirements, but they also play a crucial role in continuous improvement and strengthening the quality culture within an organization. Internal audits are essential to ensure effectiveness and excellence in quality management.

The importance of internal audits can be summed up in 7 points:

1. Objective Evaluation of the Quality Management System

Internal audits provide an unbiased and objective assessment of an organization's quality management system. Internal auditors, acting independently of the areas they audit, can identify areas for improvement and opportunities to strengthen the system. This objective assessment allows the organization to understand its current state in terms of compliance with requirements and effectiveness in implementing quality processes.

2. Identification of Non-Conformities and Deviations

One of the key functions of internal audits is to identify non-conformities and deviations from established requirements. These non-conformities may include discrepancies with quality standards, internal policies, or legal and regulatory requirements. By identifying these areas of weakness, internal audits provide an opportunity to correct problems and prevent the recurrence of non-conformities in the future.

3. Continuous Improvement and Timely Correction

Internal audits are critical to continuous improvement.

By identifying non-conformities, internal auditors can recommend specific corrective and preventive actions. These actions make it possible to address underlying problems, improve processes and procedures, and strengthen the quality management system as a whole. Timely correction of non-conformities directly contributes to the operational effectiveness and efficiency of the organization.

4. Compliance with Requirements and Regulations

Internal audits ensure compliance with legal and regulatory requirements and quality standards. By identifying potential non-compliances, internal audits allow the organization to take proactive corrective action to mitigate legal and regulatory risks. This ensures that the organization operates according to best practices and recognized standards in its industry.

5. Promoting Transparency and Accountability

Internal audits promote transparency and accountability in quality management. By providing an independent assessment of the management system, internal audits foster accountability at all levels of the organization. The results of the audits allow senior management to make informed decisions aimed at continuous improvement.

6. Preparation for External Audits and Certifications

Internal audits prepare the organization for external certification audits, such as those conducted by ISO certification bodies. By maintaining a robust and well-documented quality management system through regular internal audits, the organization can demonstrate its compliance with international standards and ensure an easier transition during external audits.

7. Strengthening the Culture of Quality

Finally, internal audits contribute to strengthening the culture of quality within an organization. By emphasizing the importance of adhering to established quality standards and requirements, internal audits promote a mindset geared toward continuous improvement throughout the organization. This creates an environment where quality is considered a shared responsibility and a priority in all organizational activities and processes.

Internal audits are essential for effective quality management, as they provide an objective assessment, identify areas for improvement, ensure compliance with requirements, and promote a quality culture geared towards continuous improvement. By implementing internal audits systematically and effectively, organizations can strengthen their management systems and achieve higher levels of performance and customer satisfaction.

3 NORMATIVE AND CONCEPTUAL FRAMEWORK

LEGAL AND REGULATORY REQUIREMENTS FOR INTERNAL AUDITS

Internal audits, especially when related to quality, environmental, safety or any other system management systems, must be carried out in accordance with a number of legal and regulatory requirements. These requirements ensure the effectiveness and integrity of the audit process, as well as compliance with internationally recognized standards.

A clear and robust understanding of the regulatory and conceptual framework is essential to conduct effective and reliable audits. In addition, it ensures that organizations derive maximum value from audits by continuously improving their management systems and operational processes.

Here are some of the main requirements to consider:

1. **Applicable Auditing Standards:**

Internal audits must follow recognized international standards and guidelines, such as ISO 19011:2018. This standard provides guidance on audit principles, the management of audit programs, and the conduct of internal and external audits. Adhering to these standards ensures consistency, effectiveness, and quality in the execution of audits.

2. **Management System Requirements:**

Internal audits must be aligned with the requirements of the applicable management system, such as ISO 9001 (Quality Management), ISO 14001 (Environmental Management), ISO 45001 (Occupational Health and Safety Management), among others. This involves verifying compliance with documented procedures and specific system requirements, ensuring that the system is implemented and maintained in accordance with established standards.

3. Legal & Regulatory Compliance:

During internal audits, compliance with legal and regulatory requirements relevant to the organization's management system and activities must be verified. This includes laws and regulations related to quality, the environment, occupational health and safety, among other aspects. Identifying legal non-conformities is critical to avoid legal risks and ensure compliance with legal obligations.

4. Ethical and Professional Standards:

Internal auditors must adhere to ethical and professional standards, such as impartiality, objectivity, integrity, and confidentiality. These standards ensure that audits are conducted fairly, transparently, and without conflicts of interest. Auditors must maintain the confidentiality of information obtained during audits and act with integrity at all times.

5. Reporting and Documentation Requirements:

It is critical to properly document all audit activities, including planning, execution, findings, conclusions, and corrective actions. Audit reports must be clear, accurate and provide sufficient evidence of the activities carried out and the results obtained. Complete and accurate documentation is essential to support the objectivity and integrity of audits.

6. Auditor Competency and Training:

Internal auditors must possess the necessary competence and training to conduct effective audits. This includes technical knowledge about the audited management system, as well as skills in auditing techniques, effective communication, problem solving, and decision-making. Ongoing training is key to maintaining and improving auditors' skills.

7. Follow-up and Review of the Audit Program:

Regular monitoring and review of the audit program is required to ensure its effectiveness and continued relevance. This involves reviewing and adjusting the program as needed based on audit results, changes in the organizational context, and updated regulatory requirements. The audit program must be flexible and adapt to the changing needs of the organization.

Complying with these legal and regulatory requirements is essential to ensure the quality, integrity and effectiveness of internal audits in the context of management systems. Compliance with these standards ensures that audits are conducted in a professional, ethical, and objective manner, thus contributing to the achievement of quality objectives and the overall success of the organization.

KEY CONCEPTS IN AUDITS AND QUALITY MANAGEMENT

Auditing and quality management are critical areas for ensuring compliance with standards, continuous improvement, and organizational success. The key concepts that provide a solid understanding of these fields are as follows:

➢ **Audit:**

Auditing is a systematic and documented process of obtaining objective audit evidence and evaluating it in an unbiased manner to determine the extent to which audit criteria are met. Auditing can cover different areas within an organization, including quality, environment, occupational health and safety, among others. Its main objective is to identify opportunities for improvement and ensure compliance with established requirements.

➢ **Internal Auditor:**

An internal auditor is a trained professional within the organization responsible for conducting internal audits. Their role is to evaluate the effectiveness of the management system and provide recommendations for continuous improvement. Internal auditors must be impartial, competent, and act in accordance with ethical and professional standards.

➢ **External Auditor:**

An external auditor is an independent expert hired by an organization to conduct external audits, usually for certification or compliance purposes. External auditors must be qualified and free from any conflicts of interest to ensure objectivity in their assessments.

> **ISO 19011:2018:**

ISO 19011:2018 provides detailed guidelines for the audit of management systems, including audit principles, management of audit programs, and conduct of internal and external audits. This standard sets out the requirements for auditors and defines key terms related to auditing, ensuring consistency and quality in audit practices.

> **Quality Management:**

Quality management refers to coordinated activities aimed at directing and controlling an organization with regard to quality. This involves establishing quality policies, goals, processes, and procedures to ensure that products or services meet customer requirements and expectations.

> **Quality Management System (QMS):**

A quality management system is a framework of policies, processes, and procedures used by an organization to plan and control its quality-related activities. A QMS is based on principles such as customer focus, leadership, staff involvement, process-based approach, and continuous improvement.

> **Non-Conformity:**

A non-conformity is a failure to comply with a set requirement, such as an ISO standard, internal procedure, or technical specification. Non-conformities can arise during an audit and require corrective actions to resolve identified deficiencies and prevent their recurrence.

> **Corrective and Improvement Action:**

Corrective actions are actions taken to eliminate the causes of a non-conformance identified during an audit or in other quality control processes. Improvement actions are measures taken to prevent the occurrence of non-conformities in the future, addressing potential causes.

> **Continuous Improvement:**

Continuous improvement is a key principle in quality management that involves the continuous effort to improve an organization's products, services, and processes. It relies on data collection and analysis, customer feedback, and the implementation of corrective and preventive actions to optimize performance and effectiveness.

> **Audit Evidence:**

Audit evidence is any verifiable information used to support the conclusions and findings of an audit. It may include documents, records, observations, interviews, and any other relevant evidence obtained during the audit process.

These concepts are critical to understanding and applying the principles of auditing and quality management effectively. Proper implementation of these concepts helps ensure compliance with requirements, early identification of problems, and continuous improvement in organizations, thus contributing to success and excellence in the quality of products and services.

4 INTERNAL AUDIT PLANNING

REQUIREMENTS AND CRITERIA FOR AUDIT PLANNING

Audit planning is a fundamental process in the systematic and objective evaluation of management systems within an organization. To ensure that audits are effective, efficient and provide meaningful results, it is crucial to follow robust planning requirements and criteria, such as those set out in ISO 19011:2018.

The requirements and criteria for audit planning encompass a number of essential considerations that must be addressed to ensure the success of this process. From determining the scope and objectives to selecting competent auditors and preparing the detailed audit plan, each step plays a crucial role in the quality and usefulness of audits.

By adhering to these requirements and criteria, organizations can maximize the value of audits by identifying areas for improvement, verifying compliance with applicable standards and regulations, and supporting the continuous improvement of their management systems. In addition, proper planning helps ensure the impartiality, reliability and professionalism of audits, thus strengthening the credibility of the process and its results.

In this context, we will look in greater detail at the key requirements and criteria for audit planning according to ISO 19011:2018, highlighting their importance and applicability in the context of modern audit practices and stakeholder expectations.

1. Determining the Scope and Objective of the Audit

Identification of the Management System: It is crucial to clearly define the management system to be audited, including its processes, functional areas and any applicable standards.

Goal Setting: Audit objectives should be clear and aligned with the organization's strategic goals. The objectives will guide the planning and execution of the audit.

Definition of Criteria and Methods: Specific criteria should be established to assess the effectiveness and conformity of the management system. Audit methods, such as interviews, document review, and observation, should also be defined.

2. Consideration of Relevant Factors

Characteristics of the Management System: The complexities, dimensions and structures of the management system must be considered in order to properly adapt the audit planning.

Significant Changes: If there have been significant changes to the system since the last audit, these should be taken into account to adjust the planning and approach of the audit.

3. Selection of Competent Auditors

Skills and Knowledge: Auditors must possess adequate technical competencies, experience, and abilities to evaluate the management system in question.

Ongoing Training: It is essential to ensure that auditors are up-to-date in their knowledge and skills relevant to the audit.

4. Preparation of the Audit Plan

Detailed Planning: The audit plan should be meticulous, including the sequence of activities, resources needed, dates, and locations of audits.

Identification of Key Areas: Critical or high-risk areas within the management system should be prioritized for special attention during the audit.

5. Communication and Confirmation of the Audit

Effective Communication: All relevant stakeholders should be informed in a clear and timely manner about the audit planning.

Formal Confirmation: The organization under audit must be formally notified of the audit, including the scope and objectives.

6. Audit Resources

Adequate Allocation of Resources: Sufficient resources (time, staff, tools) must be allocated to conduct the audit effectively and thoroughly.

7. Collection of Information

Relevant Documentation: All necessary information should be collected to support the audit, including relevant documents, records, and data.

8. Review and Approval of the Audit Plan

Stakeholder Validation: The audit plan must be reviewed and approved by all relevant stakeholders prior to implementation.

9. Plan Adjustments and Update

Flexibility and Adaptability: The audit plan should be flexible to allow for adjustments in case of unexpected changes or new circumstances that affect the audit.

These requirements and criteria ensure that audit planning is comprehensive and effective, allowing organizations to properly assess the compliance and effectiveness of their management systems against established standards and goals. The ISO 19011:2018 standard offers a solid structure for carrying out quality audits that generate meaningful and useful results for the continuous improvement of management systems.

DEVELOPMENT OF AN INTERNAL AUDIT PROGRAM

The development of an internal audit program is a critical component of effective quality and compliance management in organizations. This program establishes a systematic, planned approach to assessing compliance with standards, identifying areas for improvement, and promoting operational excellence within the organization.

The primary purpose of an internal audit program is to provide an independent and objective assessment of an organization's processes, systems, and practices. By implementing internal audits in a regular and structured manner, organizations can identify opportunities to optimize their operations, mitigate risks, and ensure compliance with applicable legal and regulatory requirements.

During the development of an internal audit program, the objectives, scope, methodologies, and timelines necessary to conduct effective audits are established. A team of competent auditors who act in an impartial and professional manner are selected and trained to carry out the assessments.

In addition, the development of an internal audit program involves the proper allocation of resources, effective communication with stakeholders, and the implementation of follow-up and continuous improvement processes based on audit findings.

A well-developed internal audit program is critical to strengthening the culture of quality and management within an organization. It provides a key tool for informed decision-making, continuous improvement, and ensuring compliance with established standards and requirements, thus contributing to the sustainable success of the organization as a whole.

Below are the key steps and considerations for establishing an internal audit program in accordance with ISO 19011:2018 guidelines.

1. **Goal Setting**

   The first step in developing an internal audit program is to set clear and specific objectives. These goals should align with the organization's overall strategy and goals. Common objectives include improving compliance with regulatory requirements, identifying areas for improvement, and promoting good management practices.

2. **Definition of Scope**

   It is essential to define the scope of the internal audit program. This includes determining which areas, processes, or systems will be audited, as well as the applicable audit criteria and standards. The scope should be realistic and encompass aspects relevant to the effective management of the organization.

3. **Selection and Training of Auditors**

   The next step is to select and train internal auditors. Auditors must possess the necessary technical competencies, as well as communication skills and the ability to act impartially. Ongoing training ensures auditors are up-to-date on relevant audit methodologies and regulatory requirements.

4. **Development of the Audit Plan**

   A detailed audit plan should be developed that includes the frequency of audits, the processes to be audited, the resources needed, and the audit criteria. The plan should be flexible to adapt to changes in the organizational environment and management priorities.

5. **Execution of Audits**

The execution of internal audits involves the performance of planned activities according to the established schedule. During the audit, auditors collect objective evidence, identify findings, and assess compliance with established requirements.

6. **Communication of Results and Follow-up**

Once the audits are completed, the results should be communicated to the relevant stakeholders. Corrective and preventive actions are identified to address non-conformities and opportunities for improvement. Follow-up ensures the effective implementation of the agreed actions.

7. **Review and Continuous Improvement**

Finally, the internal audit program should undergo periodic reviews to assess its effectiveness and efficiency. Opportunities for improvement are identified in the program itself, as well as in the audited systems and processes, to ensure continuous improvement of organizational performance.

The development of an internal audit program based on ISO 19011:2018 is critical to ensuring the effectiveness and reliability of audits within an organization. By following the steps and considerations described, organizations can establish robust internal audit programs that contribute to continuous improvement and the achievement of strategic objectives.

5 PREPARATION AND EXECUTION OF AUDITS

SELECTION AND TRAINING OF INTERNAL AUDITORS

Audit preparation and execution are critical processes that ensure the effective and objective evaluation of management systems within an organization. These processes are critical for identifying opportunities for improvement, verifying compliance with standards and requirements, and promoting operational excellence.

During the preparation of an audit, the audit objectives, scopes and criteria are established, as well as the audit team is selected and trained. Detailed planning is essential to ensure that the audit is effective and efficient, minimizing disruptions to the organization's day-to-day operations.

Audit execution involves gathering evidence, evaluating processes and practices, and identifying findings. Auditors must act with impartiality and professionalism, applying recognized audit methodologies to obtain objective and reliable results.

Proper selection and training of internal auditors is critical to ensuring the effectiveness and impartiality of audits within an organization. In this chapter, the key processes and considerations for selecting and training competent internal auditors will be addressed, supported by illustrative examples.

1. **Selection of Internal Auditors**

The selection of internal auditors should be based on specific criteria to ensure the suitability and impartiality of the candidates. Some examples of selection criteria include:

Technical Knowledge: Auditors must possess solid knowledge in the area or system to be audited. For example, an ISO 9001 internal auditor should have experience in quality management.

Communication Skills: The ability to communicate clearly and handle effective interactions is crucial for an auditor. For example, being able to interview staff and document findings.

Analytical Ability: Auditors must be able to analyze data, identify trends, and assess compliance with specific requirements.

Independence and Objectivity: Internal auditors must act impartially and not be directly involved in the areas they audit.

Example: Selection of Internal Auditors

In a manufacturing company, internal auditors are selected for the occupational health and safety management system (ISO 45001). Employees with experience in occupational safety, effective communication skills, and an in-depth understanding of the requirements of ISO 45001 are chosen.

2. Education and Training of Internal Auditors

After selection, it is crucial to provide adequate education and training to internal auditors. Some important topics to address during the training include:

Standards and Requirements: Auditors must understand the specific requirements of the management systems they audit.

Audit Methodologies: Auditors must learn auditing techniques, including evidence gathering, interviews, and reporting.

Interpersonal Skills: Training should include the development of communication, negotiation, and conflict management skills.

Ethical Practices: Auditors must understand the importance of acting ethically and maintaining confidentiality during audits.

Example: Internal Auditor Training

In a financial services company, internal auditors are trained on ISO 27001 (information security). They are trained in information security auditing techniques, handling sensitive data, and cyber risk assessment.

Careful selection and effective training of internal auditors are key elements for successful internal audits. By investing in the development of auditors' skills and knowledge, organizations can ensure high-quality audits that contribute to continuous improvement and the achievement of management objectives.

AUDIT EXECUTION PROCESS: INITIATION, PREPARATION AND IMPLEMENTATION

The process of executing audits is crucial to conducting effective and systematic evaluations of management systems in an organization. This chapter explores the key stages of the audit process, from initiation to completion, with a focus on detailed planning and impartial execution of internal audits.

1. Initiation of the Audit

Setting Audit Objectives

Before an audit is initiated, the objectives and scope of the audit should be clearly stated. Objectives may include verifying compliance with specific standards, identifying areas for improvement, or evaluating the overall performance of a management system.

Selection of the Audit Team

A team of competent internal auditors is selected, taking into account their technical knowledge, communication skills and relevant experience in the area to be audited.

Example: Initiating the Audit

In a manufacturing company, an internal audit of the quality management system (ISO 9001) is initiated with the aim of evaluating the effectiveness of production processes and meeting the requirements of the standard.

2. Preparation of the Audit

Detailed Planning

A detailed audit plan is developed that includes the objectives, scope, audit criteria, timeline, resources required, and responsibilities of the audit team.

Documentation Review

The audit team reviews relevant documentation, such as procedure manuals, quality records, and other documents related to the system to be audited.

Example: Audit Preparation

Before conducting an information security audit (ISO 27001), the audit team reviews the information security policy, security incident logs, and controls in place.

3. Conducting the Audit

Kick-Off Meetings

Initial meetings are held with audited staff to explain the purpose and scope of the audit, as well as to establish open and collaborative communication.

Evidence Gathering

The audit team collects objective evidence through observation, interviews, and document review to assess compliance with the audit criteria.

Identification of Findings

During the audit, non-conformities, areas for improvement and strengths of the management system are identified, properly documenting all findings.

Example: Conducting the Audit

During an environmental audit (ISO 14001), the audit team conducts on-site inspections, interviews with operations personnel, and reviews environmental compliance records to assess the company's environmental performance.

The audit execution process requires meticulous planning, unbiased execution, and effective communication. By following these steps, organizations can gain an objective assessment of their management systems, identify opportunities for improvement, and strengthen their commitment to operational excellence and compliance with applicable standards.

6 EVIDENCE COLLECTION AND EVALUATION

TECHNIQUES AND METHODS FOR COLLECTING EVIDENCE DURING THE AUDIT

Evidence collection and evaluation are fundamental processes in audits, as they provide the basis for informed decision-making and the identification of areas for improvement in an organization's management systems. In this context, the ability of auditors to obtain and analyze objective evidence is essential to ensure the validity and impartiality of audit findings.

During evidence collection, auditors use a variety of techniques, such as interviews, document review, direct observation, and data analysis, to gain relevant insights into compliance with requirements and the effectiveness of processes. Subsequently, the evaluation of this evidence involves critically analyzing the information collected to determine whether the audit criteria are met and to identify opportunities for improvement.

Next, the importance of evidence collection and evaluation in the context of audits will be discussed in detail, highlighting the best practices and techniques used to obtain meaningful and objective results that drive the continuous improvement of management systems.

1. Interviews

Interviews are a key technique for gathering information directly from the people involved in the audited processes. During interviews, auditors may ask specific questions to understand roles, responsibilities, practices, and perceptions about the management system.

Application Example: During a quality audit (ISO 9001), auditors interview production personnel to gain insight into standard operating procedures and quality management in the manufacturing process.

2. **Document Review**

    Document review involves analyzing records, procedures, policies, manuals, and other relevant documents related to the audited management system. This technique provides tangible evidence of implementation and compliance with requirements.

Application Example:
During an environmental audit (ISO 14001), auditors review waste management records, environmental monitoring reports, and environmental incident records to assess the organization's environmental performance.

3. **Direct Observation**

    Direct observation involves witnessing activities, processes, or conditions in the workplace. This technique allows auditors to verify the practical application of procedures and controls within the management system.

Application Example: During an occupational health and safety audit (ISO 45001), auditors look at on-site safety conditions, availability of personal protective equipment, and compliance with safety standards.

4. **Data Analysis**

    Data analysis involves reviewing and analyzing quantitative and qualitative data relevant to the audited management system. Auditors use statistical tools and analytical techniques to identify significant trends, patterns, or discrepancies.

Application Example: During an information security audit (ISO 27001), auditors analyze security incident logs, security controls performance metrics, and vulnerability test results to evaluate the effectiveness of the information security management system.

The proper use of evidence collection techniques and methods is essential to obtain an objective and complete assessment of management systems during an audit. By applying these techniques appropriately, auditors can identify areas for improvement, verify compliance with requirements, and provide useful recommendations for continuous improvement of the organization.

OBJECTIVE EVALUATION OF EVIDENCE TO IDENTIFY NON-CONFORMITIES

During an audit, the objective evaluation of the evidence is essential to identify non-conformities and areas for improvement in the audited management systems. Next, we will see how auditors can analyze evidence in an impartial and rigorous manner to detect and document significant non-conformities.

Establishment of Audit Criteria

Before evaluating the evidence, it is crucial to establish clear audit criteria based on standards, regulatory requirements, or best practices. The criteria provide the framework for determining whether the observed practices and outcomes meet expectations.

Example: For a quality management audit (ISO 9001), criteria may include compliance with documented procedures, effectiveness of corrective and preventive actions, and customer satisfaction.

Analysis of Collected Evidence

Auditors systematically analyze the evidence collected during the audit, using techniques such as document review, interviews, and direct observation. Objective evidence is sought to support compliance or identify potential non-conformities.

Example: During an information security audit (ISO 27001), auditors review security incident logs, vulnerability test results, and evidence of implementation of security controls to assess the effectiveness of the system.

Comparison with Audit Criteria

The evidence collected is compared with the established criteria to determine whether there is conformity or discrepancy. If the observed practices do not meet the defined criteria, a non-conformity is identified.

Example: If during an environmental audit (ISO 14001) it is found that adequate controls for hazardous waste management have not been put in place, this represents a non-compliance with applicable legal and regulatory requirements.

Documentation of Non-Conformities

Identified non-conformities are documented clearly and accurately, specifying the nature of the non-compliance, the evidence supporting the non-conformity, and any potential impact on the management system or operational results.

Example: A non-conformity related to the lack of training of personnel in occupational health and safety (ISO 45001) is documented, identifying the risk of occupational accidents due to lack of knowledge about safety procedures.

Objective evaluation of evidence during an audit is essential to identify significant non-conformities that require corrective and preventive actions. By following a rigorous process of analysis and comparison against established audit criteria, auditors can provide valuable recommendations to improve the effectiveness and compliance of management systems in an organization.

7 IDENTIFICATION AND CLASSIFICATION OF NON-CONFORMITIES

DEFINITION AND TYPE OF NON-CONFORMITIES

The identification and classification of non-conformities are critical processes in audits that allow the detection of significant deviations from the requirements established in the management systems. These non-conformities can range from minor non-compliances to major failures that affect the effectiveness and efficiency of organizational processes.

During an audit, the identification of non-conformities is based on the objective comparison of the evidence collected with the defined audit criteria. The identified non-conformities are classified according to their severity and their potential impact on the management system, allowing corrective and preventive actions to be prioritized to address them effectively.

Below, we will explore in detail how auditors should identify, classify, and document non-conformances during audits, highlighting the importance of these processes for continuous improvement and compliance with applicable standards and requirements in organizations.

1. Definition of Non-Conformity

A non-conformance is defined as any situation or finding that does not meet the stated requirements. It can be a lack of compliance with specific standards, documented procedures, legal/regulatory requirements, or quality expectations.

2. Types of Non-Conformities

a) Minor Non-Conformities

Minor non-conformities are deviations that do not significantly affect the effectiveness of the management system or compliance with requirements. They can be opportunities for improvement that do not pose a critical risk to the organization.

Example: Using an incorrect document format that does not affect the integrity of the content.

b) Major Non-Conformities

Major non-conformances are significant deviations that compromise the effectiveness of the management system or the ability to meet critical requirements. They require immediate corrective actions to resolve the identified issues.

Example: Lack of staff training in critical safety procedures that puts health and safety at risk in the workplace.

c) No Critical Conformities

Critical nonconformities are serious deviations that put the integrity of the management system or the safety of people, the environment, or the organization's assets at risk. They require urgent corrective actions and immediate corrective actions.

Example: Absence of safety controls in a production plant that poses an unacceptable risk to the health and safety of workers.

3. Characteristics of Non-Conformities

Non-conformities must be identifiable, clear, specific, and adequately documented to facilitate their understanding and address. They should include detailed information about the nature of the non-compliance, the evidence collected, and any potential impact on the management system.

Proper identification and classification of non-conformities during an audit are essential to improve the effectiveness and performance of management systems. By understanding the types and characteristics of nonconformities, organizations can prioritize corrective and improvement actions to address identified issues and promote continuous improvement at all levels of the organization.

CRITERIA FOR CLASSIFYING AND PRIORITIZING THE NON-CONFORMITIES FOUND

Proper classification and prioritization of non-conformances found during an audit are essential to ensure that resources are allocated effectively and critical areas impacting the effectiveness and compliance of management systems are addressed. Now, you will look at the criteria used to classify and prioritize non-conformities, along with examples that illustrate their practical application in specific contexts.

1. Severity of Impact

The severity of the impact of a non-conformity is one of the main criteria for its classification and prioritization. The level of negative effect that non-conformity may have on processes, products, services or stakeholders is assessed. Typical levels of severity include:

Minor Non-Conformance: A slight impact that does not significantly affect the effectiveness of the management system or compliance with requirements.

Example: Incorrect use of a document format that does not affect the integrity of the content.

Major Non-Conformity: Moderate impact that partially compromises the effectiveness of the management system or compliance with requirements.

Example: Lack of staff training in quality-critical procedures that can affect the quality of the final product.

Critical Non-Conformity: Severe impact that seriously compromises the effectiveness of the management system or poses a significant risk to safety, health, or the environment.

Example: Absence of safety controls in a production plant that poses an unacceptable risk to the health and safety of workers.

2. Frequency or Recurrence

The frequency or recurrence of a non-conformance can influence its prioritization. The most frequently occurring non-conformities may indicate systemic issues that need to be addressed on a priority basis to prevent future non-compliances.

Example: If a non-conformance related to a lack of equipment maintenance records is identified in multiple areas of the organization, this may indicate a deficiency in maintenance procedures that needs to be addressed urgently to prevent future failures.

3. Potential Risk

The level of risk associated with a non-conformance is also a key criterion for prioritization. Non-conformities that pose a significant risk to safety, health, the environment, or legal compliance should receive priority attention and immediate corrective action.

Example: A non-conformity related to non-compliance with environmental regulations that may result in legal penalties or adverse environmental impacts should be treated as a high priority to mitigate legal risks and protect the organization's reputation.

4. Impact on Customers or Stakeholders

The impact on customers or stakeholders can also be considered when prioritizing non-conformities. Those that directly affect customer satisfaction or stakeholder expectations may require immediate attention to maintain the organization's trust and reputation.

Example: A non-conformity related to late delivery of products that affects customer satisfaction and business relationships may be prioritized to restore customer trust and prevent business losses.

By using clear and objective criteria to classify and prioritize non-conformances found during an audit, organizations can effectively manage risks, improve the effectiveness of their management systems, and ensure compliance with regulatory requirements and quality expectations. Timely identification and attention to non-conformities contributes significantly to continuous improvement and organizational success. Proper prioritization allows you to focus efforts on critical areas that have the greatest impact on your organization's overall quality, safety, and performance.

8 CORRECTIVE AND IMPROVEMENT ACTIONS

DEVELOPMENT OF CORRECTIVE AND IMPROVEMENT ACTIONS

Corrective and improvement actions are fundamental processes within a management system that address non-conformities, solve identified problems, and promote continuous improvement in an organization. These actions are key to correcting deficiencies, preventing the recurrence of problems, and optimizing performance at all levels.

In this context, corrective actions focus on eliminating the root causes of non-conformances identified during an audit or assessment process. On the other hand, improvement actions seek to implement proactive changes aimed at optimizing processes, products or services and increasing the effectiveness of the management system.

Next, we will look in detail at the concept and importance of corrective and improvement actions in the context of quality management and management systems in general. We will discuss the key steps to implement effective actions that drive operational excellence and foster a culture of continuous improvement in organizations.

1. **Identification and Documentation of Non-Conformities**

    The first step in developing corrective and improvement actions is to properly identify and document nonconformities. This can arise from a variety of sources, such as internal audits, customer feedback, data analysis, or performance reviews.

Information Gathering: Gathers evidence, data, and observations related to the identified non-conformity.

Detailed Documentation: Clearly outlines the nature of the non-conformity, including its impact, affected areas, and any relevant information.

Example: During a process review, a non-conformance related to delays in product delivery due to inventory management issues is identified.

2. Root Cause Analysis

Once the non-conformance is identified, a root cause analysis is performed to understand why it occurred and what factors contributed to it. The goal is to identify the underlying causes that need to be addressed to prevent recurrence of the problem.

Analysis Tools: Use techniques such as the 5 Whys, Ishikawa diagrams (fishbone), FMEA (Failure Mode, Effect and Criticality Analysis) analysis, or other appropriate tools.

Identification of Contributing Factors: Identifies the factors or conditions that have contributed to the occurrence of the nonconformity.

Example: Root cause analysis reveals that delivery delays are due to demand forecasting issues and ineffective inventory management processes.

3. Development of Corrective and Improvement Actions

Based on root cause analysis, targeted corrective and improvement actions are developed to address the underlying causes identified and prevent recurrence of the problem.

Corrective Actions: Establishes actions to correct the root causes of the identified nonconformity.

Improvement Actions: Proposes proactive actions to improve existing processes, systems, or practices and prevent future problems.

Example: Corrective actions are developed that include reviewing and improving demand forecasting processes, as well as implementing a more efficient inventory management system.

4. Implementation and Follow-up

The corrective and improvement actions developed are implemented in the organization, and continuous monitoring is carried out to ensure their effectiveness and make adjustments if necessary.

Efficient Implementation: Implements planned actions in a timely and effective manner, i.e. making a contribution to the management system.

Results Monitoring: Follow up regularly to evaluate progress and verify the effectiveness of the actions implemented.

Example: The actions developed are implemented in the operations department, and a follow-up process is established to evaluate the improvement in delivery times and inventory efficiency.

The proper development of corrective and improvement actions is essential to promote continuous improvement and strengthen management systems in organizations. By following a structured and systematic approach, organizations can effectively address nonconformities, resolve fundamental issues, and optimize their operations to achieve higher levels of performance and customer satisfaction. The successful implementation of these actions is key to cultivating an organizational culture oriented towards quality and operational excellence.

EFFECTIVE IMPLEMENTATION OF ACTIONS TO RESOLVE NON-CONFORMITIES

In any organizational environment, non-conformities are critical occasions that require an effective and structured response to ensure continuous improvement and quality of processes, products, or services. A non-conformance can arise from a variety of sources, such as errors in processes, deficiencies in staff training, equipment failures, or discrepancies with established requirements.

Implementing appropriate corrective actions is critical to effectively address non-conformities and prevent their recurrence in the future. This process involves not only correcting the immediate problem, but also identifying and addressing the underlying causes or those that are not obvious, to promote a culture of continuous improvement in the organization.

The key elements for the effective implementation of corrective actions will be discussed, from the initial assessment of non-conformities to monitoring and post-implementation evaluation. Through a structured and systematic approach, organizations can transform the challenges of non-conformances into opportunities for meaningful improvement, thereby strengthening their ability to deliver high-quality products and services and maintain customer satisfaction.

1. Detailed Non-Conformity Assessment

Detailed assessment of non-conformances is the crucial first step in implementing effective corrective actions. This process not only involves identifying discrepancies or failures in processes, products, or services, but also thoroughly understanding the underlying causes that contributed to these nonconformities.

➢ Root Cause Analysis

Root cause analysis is a critical tool for identifying the true reasons behind non-conformities. Methods such as the Ishikawa Diagram (also known as the fishbone diagram) or the "5 Whys" method are useful in this context. These approaches help to dig deeper beyond superficial symptoms and identify the root causes that triggered nonconformities.

Identification of the Problem or Non-Conformity: The first step is to clearly define the problem or non-conformity that is being addressed. This can be any discrepancy between the current performance and the expected standard in a management system, such as repeated errors, quality failures, process delays, among others.

Data Collection: Relevant data related to the issue is collected, such as incident logs, quality reports, performance data, staff observations, etc. It is important to have accurate and complete information in order to conduct a thorough analysis.

Data Analysis: Specific tools and techniques are used to analyze the data collected and detect patterns, relationships, or trends that may be contributing to the problem. Some common tools include:

- **Ishikawa Diagram (Cause and Effect or Herringbone Diagram):** Organizes the possible causes of a problem into categories such as methods, labor, materials, machines, environment, and measurements. It helps to visualize and understand the interrelationships between the different causes.

- **5 Why:** This approach involves repeatedly asking the question "Why?" to dig deeper into the underlying causes of the problem. This process is continued until you reach the root cause.

- **Pareto Analysis:** Identifies and prioritizes the most significant causes contributing to the problem based on their frequency or impact.

Root Cause Identification: After analyzing the data using the appropriate tools, the root cause or root causes that explain why the problem occurred is identified. The root cause is the most basic or fundamental cause that, if left unaddressed, can result in the recurrence of the problem.

Development of Corrective Actions: Once the root causes are identified, specific corrective actions are developed to address each of them. These actions should be designed to eliminate or mitigate the root causes and prevent recurrence of the problem.

➢ **Impact & Reach**

It is essential to assess the current and potential impact of non-conformities on the organization. This involves considering how they affect customers, product or service quality, operational efficiency, and the organization's reputation. Understanding the full extent of non-conformances provides a solid foundation for developing an effective plan of action.

Example 1: Non-Conformity in a Production Process

Description of Nonconformity: Recurring failures in a production equipment that cause disruptions in the manufacturing line.

Impact Assessment:

Operational Impact: Halts production, affecting the ability to fulfill orders and delivery deadlines.

Product Quality Impact: Can lead to defective or out-of-specification products, impacting customer satisfaction.

Cost Impact: Increased costs due to downtime, frequent repairs, and rework.

Scoping Assessment:

Departments Affected: Production, Quality Control, Maintenance.

Impacted Products or Production Lines: All products that pass through the affected line.

Affected Customers: Customers who are waiting for deliveries or who have received defective products.

Example 2: Non-Compliance in an Administrative Process

Description of Non-Conformance: Errors in data entry into an inventory management system.

Impact Assessment:

Impact on Inventory Accuracy: It can result in discrepancies between physical and registered inventory, affecting product availability.

Impact on Decision Making: Inaccurate data can lead to erroneous decisions in production planning or purchasing.

Impact on Profitability: Loss of revenue from lost sales or additional costs due to incorrect inventories.

Scoping Assessment:

Functional areas affected: warehouse, purchasing, sales.
Impacted Systems or Tools: Inventory Management System, Accounting Software.
Affected Stakeholders: Warehouse Staff, Operations Management, Accounting.

Evaluation Methods:

To assess the impact and scope of a non-conformity, methods such as:

Interviews and Consultations: Talk to the staff involved to understand how non-conformity affects their day-to-day activities and areas of responsibility.

Data Analysis: Review operational records, financial reports, and performance data to quantify impact in numerical terms.

Customer and Supplier Analysis: Obtain feedback from affected customers or related suppliers to understand how non-conformance impacts their experience or relationship with the organization.

Risk Analysis: Identify potential additional risks that could arise as a result of non-conformity.

2. Development of a Corrective Action Plan

Once the root causes have been identified and the impact assessed, it's time to develop a detailed, results-oriented corrective action plan.

> **Measurable Goals**

Corrective action goals should be specific, measurable, achievable, relevant, and time-bound (known as SMART goals). These objectives clearly state what is expected to be achieved through the implementation of corrective actions.

Specific: Objectives should be clear and detailed, focused on a specific and well-defined outcome. It is important to avoid ambiguity and ensure that everyone has a clear understanding of what needs to be achieved.

Measurable: Goals must be quantifiable so that they can be objectively evaluated. They should include indicators or metrics that allow progress to be measured and determine whether the goal has been achieved.

Achievable: Goals should be realistic and achievable within available resources and constraints. They must be challenging, but they must be feasible with effort and commitment.

Relevant: Objectives must be aligned with the organization's vision and mission, as well as its strategic priorities. They should contribute directly to resolving identified non-conformities and improving the overall performance of the management system.

Time-bound: Objectives must have a defined time frame for their achievement. Setting a deadline helps maintain focus and discipline, and allows you to evaluate progress effectively.

Let's look at how to apply SMART principles in a management system when setting goals to implement corrective actions:

Example of Application of SMART Objectives in a Management System:

No SMART: "Reduce non-conformities in the production department."

This goal is vague and doesn't meet SMART principles because it doesn't specify how the reduction will be measured, whether it's achievable within a given timeframe, or how it aligns with the organization's overall strategy.

SMART: "Reduce product quality-related non-conformities in the production department by 50% by the end of the third quarter, by implementing additional quality control trainings and introducing more rigorous verification procedures."

Specific: The type of non-conformities to be addressed (related to product quality) and the specific actions to be taken (additional trainings, verification procedures) are clearly identified.
Measurable: A quantitative objective is established (to reduce non-conformities by 50%) that allows progress to be evaluated objectively.
Achievable: Although ambitious, the goal is realistic considering the implementation of additional trainings and procedures.
Relevant: The reduction of non-conformities is directly aligned with the objective of improving product quality in the production department.
Time-bound: A clear deadline (end of the third quarter) is set for achieving the goal, providing a time frame to evaluate progress and results.

➤ Detailed Activities

The action plan should include a detailed list of specific activities that will be carried out to address each root cause identified. Each activity should be clearly defined, with clear steps and assignment of responsibilities.

➤ Responsibilities and Deadlines

Assign clear responsibilities to the people or teams in charge of executing each activity. Setting realistic timelines for each stage of the plan ensures steady progress and timely implementation of corrective actions.

3. Allocation of Resources and Support

Effective implementation of corrective actions requires adequate resources and strong organizational support.

➤ **Resource Allocation**

Ensure the availability of the financial, technological, and human resources necessary to execute corrective actions effectively. This may include allocated budgets, access to specific tools and technologies, and the allocation of trained staff.

➤ **Education & Training**

If necessary, provide additional training to personnel involved in the implementation of corrective actions. This ensures that everyone understands their role and responsibilities, as well as best practices for addressing non-conformities.

4. Continuous Monitoring and Follow-Up

Once corrective actions are underway, it is essential to maintain constant monitoring to ensure that the expected results are being achieved.

➤ **Periodic Reviews**

Schedule regular reviews to assess the progress of the action plan. During these reviews, emerging challenges should be identified and the plan adjusted as needed to ensure continued effectiveness.

➤ **Key Performance Indicators (KPIs)**

Establish key performance indicators to measure the success of corrective actions. These KPIs can include reducing defects, improving customer satisfaction, or operational efficiency.

An example of a Key Performance Indicator (KPI) related to quality management could be the "Non-Conformance Index". This KPI is used to measure the effectiveness of a management system in identifying and correcting non-conformities in a given period. Here's how it's defined and calculated:

Example of KPIs: Non-Conformance Index

Definition: The Non-Conformance Index is the percentage of non-conformities identified with respect to the total number of activities or processes evaluated in a given period of time.

KPI formula:

$$\text{No Conformities Index} = (\text{Number of No Conformities Identified} / \text{Total of Evaluated Activities or Processes}) \times 100\%$$

Application Example:

Let's assume that in one month 100 internal quality audits were conducted in a company and a total of 20 non-conformities were identified during these audits.

Number of Node Conformities Identified = 20
Total Activities or Processes Evaluated = 100

Substituting these values into the KPI formula:

Non-Conformities index = 20%

Therefore, the Non-Conformance Index for that month is 20%. This means that 20% of the activities or processes assessed during internal audits presented non-conformities that required corrective actions.

A lower Non-Conformance Index (e.g., close to 0%) indicates a good performance of the management system, where few non-conformities are identified in relation to the total number of activities evaluated. On the other hand, a higher Non-Conformance Index (e.g., more than 10-15%) may indicate potential areas for improvement in the management system, such as the need to strengthen controls or staff training.

5. Communication and Feedback

Open communication and feedback are critical to keeping all stakeholders informed and engaged during the implementation process.

➢ Effective Communication

Regularly report to all stakeholders on the progress and results of corrective actions. This fosters transparency and maintains organizational support. Always remember to generate logs, that's why emails are recommended.

➢ Collecting Feedback

Solicit feedback from stakeholders and other relevant stakeholders to identify areas for improvement or necessary adjustments in the implementation process.

- **Feedback Surveys**

Surveys allow you to collect opinions and comments in a structured and anonymous way. They can be used to assess employee satisfaction, perception of the work environment, process effectiveness, among other aspects. Tools such as SurveyMonkey, Google Forms, or even surveys integrated into HR management platforms are useful for designing and distributing feedback surveys.

- **Structured Interviews**

    Conducting one-on-one or group interviews can provide valuable insights into experiences, ideas, and concerns. Structured interviews follow a set of predefined questions that allow data to be collected consistently. This technique is commonly used in performance reviews, project evaluations, or cross-team feedback sessions.

- **360° Feedback**

    This technique involves gathering opinions about an individual from multiple sources, including supervisors, teammates, subordinates, and internal or external customers. The goal is to provide a comprehensive view of a person's performance and competencies from different perspectives. Tools like Qualtrics, Trakstar, and Culture Amp offer solutions for conducting 360° feedback assessments.

- **Feedback Meetings**

    Regular meetings dedicated to feedback allow you to openly discuss important topics, share ideas, and solve problems in real time. These meetings can be one-on-one (such as performance reviews) or group (such as team feedback sessions). Video conferencing platforms such as Zoom, Microsoft Teams, or Google Meet make it easy to hold virtual meetings for remote feedback.

- **Communication & Collaboration Tools**

    Team communication tools, such as Slack, Microsoft Teams, Asana, or Trello, make it easy for team members to communicate and collaborate. These platforms allow you to instantly exchange feedback, share files, and have structured conversations about specific projects or tasks.

- **Feedback Boards**

    Feedback boards are visual tools that allow teams to collaboratively share feedback, ideas, and suggestions. Tools like Miro, Trello, or even physical whiteboards can be used to collect and organize feedback visually, making it easier to identify trends and priorities.

6 Post-Implementation Evaluation

    Once corrective actions are complete, it is essential to evaluate the results to ensure that non-conformities have been properly addressed.

➢ **Verification of Results**

    Compare the results obtained with the defined objectives to determine if the non-conformities have been satisfactorily resolved.

➢ **Lessons Learned**

    Identify lessons learned during the implementation process and document them for future reference. These lessons can inform and improve quality management practices in the organization.

    Effective implementation of corrective actions requires a systematic approach, adequate resources, and clear, ongoing communication. By following this detailed process, organizations can proactively address non-conformances and achieve significant improvements in their processes and outcomes.

9 MANAGEMENT AND MONITORING OF NON-CONFORMITIES

NON-CONFORMANCE MANAGEMENT PROCESS

Effective management of non-conformities is an essential component in any quality management system aimed at continuous improvement. Non-conformities represent deviations from established requirements or expected quality standards, and their proper management and monitoring are essential to ensure operational excellence and customer satisfaction.

The process of managing and monitoring non-conformities will be explored, from their initial detection to their subsequent resolution and follow-up. We will discuss strategies, tools, and best practices that organizations can employ to effectively handle non-conformances and promote continuous improvement at all levels.

1. Identification of Non-Conformities

Accurate identification of non-conformances is the critical first step in the management process. Non-conformities can manifest through audits, customer complaints, quality inspections, process reviews, or other internal control mechanisms. Robust non-conformance detection and registration systems are essential.

Key Activities:

Information Collection: Obtaining data and observations related to detected non-conformities.

Detailed Record: Documentation of the nature of the non-conformity, location, date, potential impact, and persons involved.

2. Evaluation and Analysis

Once identified, non-conformities need to be assessed and analysed in depth to understand their scope and underlying causes. Root cause analysis is an effective technique for identifying the root reasons behind non-conformances and determining the necessary corrective actions.

Key Activities:

Root Cause Analysis: Use of tools such as the Ishikawa Diagram (Fishbone), the 5 Whys or Pareto analysis to identify root causes.

Impact Determination: Assessment of the impact of non-conformities in terms of product quality, customer satisfaction, operating costs, and regulatory compliance.

3. Development of Corrective and Improvement Actions

Based on the analysis performed, corrective and improvement actions are developed and implemented to effectively address non-conformities and prevent their recurrence in the future.

Key Activities:

Corrective Actions: Implementation of immediate measures to correct the identified problem and restore compliance with established requirements.

Improvement Actions: Design and implementation of improvement actions to eliminate underlying root causes and prevent future non-conformities.

4. Implementation and Follow-up

Once developed, corrective and preventive actions are implemented throughout the organization. It's crucial to continuously follow up to check the effectiveness of actions and make adjustments as needed.

Key Activities:

Assignment of Responsibilities: Designation of those responsible for the implementation of actions and establishment of clear deadlines.

Follow-up and Verification: Monitoring the progress of actions through key performance indicators (KPIs) and follow-up audits.

5. Closing and Effectiveness Verification

Once corrective and improvement actions are completed, their effectiveness is verified to ensure that non-conformities have been satisfactorily resolved.

Key Activities:

Verification of Results: Comparison of the current status with the established requirements to confirm compliance and closure of non-conformity.

Lessons Learned: Identification of opportunities for improvement and lessons learned to strengthen the quality management system.

The non-conformance management process is a continuous cycle of identification, analysis, action, and follow-up designed to improve an organization's operational quality and efficiency. By following this process in a systematic and focused manner, organizations can effectively resolve nonconformities, minimize risks, and promote a culture of continuous improvement at all levels. Effective non-conformance management is key to maintaining excellence and competitiveness in today's market.

FOLLOW-UP AND CLOSURE OF NON-CONFORMITIES TO ENSURE THE EFFECTIVENESS OF ACTIONS

Developing an effective strategy for the follow-up and closure of non-conformities is critical to ensure that the corrective and preventive actions implemented are effective and fully resolve the issues identified. Here's a detailed strategy you can implement in your organization:

1. Establish Clear Roles and Responsibilities

It is critical to designate specific and clear roles within the quality management team to ensure that responsibility for the follow-up and closure of non-conformities is clear and well understood. Some key roles may include:

Non-Conformance Officer: Responsible for coordinating non-conformance management, allocating resources, and overseeing overall progress.

Responsible Teams: Designate teams or individuals responsible for implementing specific corrective and improvement actions.

Establishing clear roles helps ensure that tasks related to non-conformance management are addressed in a timely and effective manner.

2. Define Key Performance Indicators (KPIs) for Monitoring

Choosing the right KPIs is essential to measure the success and effectiveness of the non-conformance follow-up and closure process. Some relevant KPIs may include:

Nonconformance Recurrence Rate: The percentage of nonconformities that reoccur after the implementation of corrective actions.

Average Nonconformance Closure Time: The time required to fully resolve a nonconformance from detection to closure.

Customer Satisfaction After Improvements: Evaluation of customer satisfaction after implementing corrective and improvement actions.

These KPIs provide tangible metrics to evaluate the impact and effectiveness of the actions taken.

3. Implement a Centralized Tracking System

Using a centralized non-conformance management system makes it easier to follow up and properly document each case. An effective system should enable:

Detailed Record of Non-Conformities: Complete documentation of each non-conformity, including description, causes, actions taken and responsible.

Automated Action Tracking: Monitoring the implementation status of each corrective and improvement action.

Integration with other management tools (such as quality systems or enterprise software) can improve process efficiency.

4. Establish Clear Timelines and Deadlines

Defining clear timelines for the implementation of actions is essential to maintain momentum and accountability. It's important:

Assign Realistic Deadlines: Set achievable deadlines for each corrective and preventive action.

Schedule Periodic Reviews: Schedule regular reviews to assess progress and make adjustments if necessary.

Meeting deadlines contributes significantly to the timely resolution of non-conformities.

5. Conduct Follow-Up Audits

Follow-up audits are essential to verify the effectiveness of the actions implemented and ensure ongoing compliance with quality standards. During audits:

Verify Procedural Compliance: Confirm that corrective and preventive actions were implemented as planned.

Evaluate the Effectiveness of Actions: Determine if non-conformities have been effectively resolved and have not reoccurred.

Audits provide an opportunity to identify additional areas for improvement.

6. Continuous Feedback and Communication

Maintaining open and continuous communication with the teams involved in non-conformance management is key to the success of the process. This involves:

Regular Meetings: Schedule regular meetings to discuss progress, identify challenges, and share lessons learned.

Solicit Feedback: Gather feedback on the effectiveness of the actions implemented and the improvements needed.

Feedback allows you to adjust your approach and address issues proactively.

7. Review and Improve Processes

Based on the results obtained from the monitoring and closure of non-conformities, it is important to carry out a critical review of the existing processes and procedures. This includes:

Identify Areas for Improvement: Identify opportunities to optimize non-conformance management and strengthen the quality system.

Implement Changes: Make necessary adjustments and modifications to improve the effectiveness and efficiency of the process.

Continuous improvement ensures long-term adaptability and operational excellence.

8. Document Lessons Learned

Recording and documenting lessons learned during the non-conformance tracking and closure process is essential for organizational development. This involves:

Document Best Practices: Record effective practices and recommendations for future situations.

Knowledge Sharing: Communicate lessons learned through training sessions, documents, or knowledge tools.

Organizational learning promotes innovation and resilience.

By implementing this detailed strategy for tracking and closing non-conformities, organizations will be able to significantly improve their ability to effectively manage deviations, ensure the quality of their products or services, and promote a culture of continuous improvement. Dedication to rigorous monitoring and proper closure of non-conformities is critical to maintaining excellence and competitiveness in today's marketplace.

10 FOLLOW-UP AUDITS AND SYSTEM REVIEW

IMPORTANCE OF FOLLOW-UP AUDITS IN CONTINUOUS IMPROVEMENT

System monitoring and review audits are fundamental pillars in the field of management systems, ensuring continuous effectiveness, compliance and constant improvement within an organization. In management systems, we recognize the critical importance of these audits to evaluate and optimize the performance of the systems implemented.

The main objective of follow-up audits is to evaluate the implementation and maintenance of a management system over time. They provide a key opportunity to verify the effectiveness of corrective actions taken, identify areas for improvement, and ensure continued compliance with established requirements and standards.

On the other hand, system review audits provide a comprehensive assessment of the overall effectiveness of the management system. These audits examine not only compliance with standards and regulations, but also alignment with the organization's strategic goals and stakeholder expectations.

1. Identification of Opportunities for Improvement

Early detection of deviations and non-conformities are causes of follow-up audits that allow deviations and non-conformities to be detected at an early stage, before they become major problems. This facilitates the implementation of timely and effective corrective actions, minimizing negative impacts on processes, products, or services.

Root Cause Analysis is a critical part of audits to identify the underlying reasons for non-conformances. This analysis provides valuable information to address root causes and prevent problems from recurring.

Follow-up audits also evaluate the effectiveness of previously implemented corrective and improvement actions. This makes it possible to determine whether the measures taken have been effective and, if not, identify the need for additional or alternative actions.

3. System Performance Evaluation

Follow-up audits verify compliance with legal, regulatory, normative, and other requirements applicable to the management system. This ensures that the organization meets external and internal obligations and expectations.

Auditors evaluate key performance indicators (KPIs) related to the audited processes, products, or services. This allows you to identify trends, make comparisons, and determine if the established objectives and goals are being achieved.

Audits reveal the strong areas of the system, where processes and controls are working effectively, as well as the weak areas that require attention and improvement. This information is crucial for prioritizing improvement efforts and allocating resources efficiently.

4. Fostering a Culture of Continuous Improvement

Follow-up audits involve staff at all levels of the organization, promoting their participation in the continuous improvement process. This fosters a sense of ownership and commitment to improving the system.

Audit findings and recommendations are communicated in a clear and transparent manner to all relevant stakeholders. This raises awareness of areas for improvement and promotes buy-in and adoption of corrective and preventative actions.

Audits also identify and recognize achievements and good practices within the organization. This reinforces positive behaviors and motivates staff to continue improving.

5. Corrective and Improvement Actions

Based on the findings of the audits, corrective actions are implemented to address and eliminate identified non-conformities. This ensures that concrete steps are taken to improve the performance of the system.

Follow-up audits also include monitoring and verifying the effectiveness of corrective and improvement actions implemented. This ensures that the measures taken are effective and the expected results are achieved.

6. System Review and Update

Information gathered during audits, including findings, nonconformities, corrective and improvement actions, is analyzed in depth to identify patterns, trends, and areas of opportunity.

Based on the analysis of audit information, opportunities for improvement in the processes, procedures and policies of the management system are identified. This can lead to changes and optimizations to improve the efficiency and effectiveness of the system.

Once opportunities for improvement have been identified, relevant components of the management system, such as manuals, procedures, and documentation, are updated and revised to reflect the changes and improvements implemented.

7. Change Management

Planning and controlling changes resulting from improvements: Implementing improvements and changes in the management system requires careful planning and control. Action plans should be established, responsibilities assigned, deadlines defined, and the progress of changes monitored.

It is essential to train and sensitize staff about the changes implemented in the management system. This ensures a proper understanding of new processes, procedures, and requirements, facilitating a smooth transition and effective adoption.

Once changes are implemented, their effectiveness needs to be monitored and evaluated. This may involve conducting specific follow-up audits to determine whether the changes have achieved the expected results and whether additional adjustments are required.

8. Organizational Learning

Follow-up audits generate valuable lessons about what worked well, challenges encountered, and best practices identified. These lessons are documented and disseminated throughout the organization to facilitate continuous learning.

An environment of feedback and knowledge sharing is promoted, where staff can share their experiences, ideas and suggestions for improvement. This enriches the organizational learning process and fosters collaboration.

Best practices identified through audits and organizational learning are integrated into the organization's processes and operations. This ensures that the knowledge gained becomes an integral part of the management system and contributes to continuous improvement.

9. Management Commitment

The commitment and leadership of senior management are critical to the success of continuous improvement driven by follow-up audits. Management should actively support improvement initiatives, promote a culture of continuous improvement, and allocate the necessary resources.

Continuous improvement requires the allocation of adequate resources, such as trained personnel, time, budget, and tools. Senior management must ensure that the necessary resources are allocated to implement and maintain an effective follow-up audit and continuous improvement program.

Management must establish clear objectives and goals for continuous improvement, aligned with organizational strategy and objectives. These objectives and goals should be communicated at all levels of the organization and serve as a guide for improvement activities.

10. Benefits of Continuous Improvement

By identifying and eliminating inefficiencies, waste, and non-conformities, follow-up audits contribute to improving the efficiency and productivity of the organization's processes and operations.

The corrective and improvement actions resulting from the audits aim to improve the quality of products and services, reducing defects, errors and ensuring compliance with established requirements.

By improving the quality and performance of processes, products, and services, organizations can better meet the expectations and needs of customers and other stakeholders, leading to greater satisfaction and loyalty.

An organization that effectively implements continuous improvement through follow-up audits can achieve a competitive advantage by offering higher quality, efficient, and valuable products and services. In addition, continuous improvement contributes to the sustainability and long-term growth of the business.

It highlights how audits identify opportunities for improvement, evaluate system performance, implement corrective actions, review and update the system, manage change, foster organizational learning, and ensure management engagement. In addition, it highlights the tangible benefits that come from integrating follow-up audits into the continuous improvement process, such as increased efficiency, quality, customer satisfaction, and competitive advantage, leading to greater sustainability and business growth.

REVIEW OF THE MANAGEMENT SYSTEM BASED ON AUDIT FINDINGS

The review of the management system is a fundamental process to ensure the effectiveness, adequacy and continuous alignment of the system with the objectives and strategies of the organization. This review relies heavily on the findings and results obtained during follow-up audits and other relevant sources of information.

Follow-up audits play a crucial role in identifying areas for improvement, non-conformities, and opportunities for optimization within the management system. These findings provide an objective and detailed view of the system's performance, its compliance with the established requirements, and its ability to achieve the objectives set.

Through the review of the management system, senior management and system managers can comprehensively analyze and evaluate the information collected during audits. This includes:

1. Non-conformance analysis: Non-conformities identified during audits are analyzed in depth to understand their root causes and determine the most appropriate corrective and improvement actions.

2. Evaluation of the effectiveness of the actions taken: The effectiveness of the corrective and improvement actions implemented previously is evaluated, verifying whether they have been successful in eliminating non-conformities and preventing their recurrence.

3. Identification of opportunities for improvement: Audit findings may reveal opportunities for improvement in processes, procedures, policies, and other aspects of the management system, which are considered during the review.

4. Trend and Pattern Analysis: By examining the results of multiple audits, it is possible to identify trends and patterns that may indicate problem areas or systematic strengths within the management system.

5. Assessment of compliance with requirements: The review of the system also involves assessing compliance with applicable legal, regulatory, normative, and other requirements, using the information gathered in audits.

6. Overall Performance Review: The overall performance of the management system is reviewed, including key performance indicators, objectives achieved, and areas requiring additional attention.

This comprehensive review of the management system, based on audit findings, allows the organization to make informed decisions and establish action plans to address identified deficiencies, implement improvements, and adjust the system as needed.

In addition, the management system review fosters continuous improvement by identifying and addressing areas of opportunity, promoting a proactive and constant learning approach. This ensures that the management system is kept up-to-date, effective, and aligned with the changing needs of the organization and its stakeholders.

Reviewing the management system based on audit findings is an essential process that allows organizations to comprehensively evaluate the performance of their system, identify areas for improvement, and take corrective and preventive actions to maintain a robust, efficient, and ever-improving management system.

Detailed Analysis of Findings

After completing an audit, it is essential to conduct a detailed analysis of the findings obtained. This analysis involves not only identifying non-conformities, observations, and areas for potential improvement, but also understanding the underlying causes of the identified issues. A thorough analysis allows findings to be prioritized based on their impact on the management system and facilitates informed decision-making on corrective and preventive actions to be implemented.

Effective Corrective and Improvement Actions

Based on audit findings, effective corrective and improvement actions should be implemented. Corrective actions focus on addressing identified non-conformities, correcting root causes to prevent their recurrence. It is crucial to design proactive actions that mitigate future risks and improve the resilience of the management system. These actions must be specific, measurable, achievable, relevant, and time-bound (SMART) to ensure their effectiveness.

Strategic Review of the Management System

Audit findings provide a solid foundation for strategically reviewing the management system. This review goes beyond addressing individual non-conformities; The aim is to assess the overall effectiveness of the system in relation to the organisation's strategic objectives. During the review, areas may be identified that require adjustments to document structure, operational processes, human resources, or the assignment of responsibilities. This critical assessment ensures that the management system is aligned with the changing needs of the organization and the environment.

Promotion of Continuous Improvement and Innovation

Effective utilization of audit findings promotes continuous improvement and fosters innovation in the organization. These findings can inspire new ideas to optimize processes, adopt emerging technologies, or develop more efficient practices. By actively integrating audit results into the organizational culture, an environment is fostered where continuous improvement becomes a priority shared by all team members.

Monitoring and Evaluation of Results

Once corrective and preventive actions have been implemented, it is essential to monitor and evaluate the results obtained. This continuous monitoring makes it possible to verify the effectiveness of the actions taken and make adjustments as needed. Monitoring also provides valuable data to measure the overall impact of improvements on management system performance and the achievement of organizational objectives.

Audit findings not only reveal weaknesses in the management system, but also offer an invaluable opportunity to drive organizational improvement and innovation. By effectively leveraging these findings through solid improvement and corrective actions, meaningful strategic reviews, and a focus on continuous improvement, organizations can strengthen their ability to adapt, promote operational excellence, and stay competitive in a dynamic business environment.

11 CONTINUOUS IMPROVEMENT AND ORGANIZATIONAL LEARNING

USING LESSONS LEARNED FROM AUDITS TO IMPROVE PROCESSES AND SYSTEMS

In a highly competitive and ever-evolving business environment, an organization's ability to continuously improve and adapt to change is critical to its long-term success and sustainability. Continuous improvement and organizational learning are closely related concepts that enable companies to make the most of their potential and stay ahead of the curve.

Continuous improvement is a systematic and proactive approach that seeks to identify and take advantage of optimization opportunities in an organization's processes, products, and services. This approach is based on the principle that there is always room for growth and improvement, regardless of the current level of performance. Continuous improvement involves a constant commitment to excellence and the search for more efficient and effective ways of doing things.

Organizational learning refers to an organization's ability to acquire, share, and apply knowledge and experiences in a systematic and continuous manner. It is a process that involves all members of the organization and encourages the acquisition, transfer and application of knowledge and best practices at all levels.

Continuous improvement and organizational learning are complementary and interdependent concepts. To achieve effective continuous improvement, the organization needs to be willing to learn from its experiences, evaluate its processes and results, and apply the knowledge gained to make adjustments and optimizations. In turn, organizational learning fosters an environment conducive to continuous improvement by promoting openness to change, experimentation, and innovation.

Some of the key practices to foster continuous improvement

and organizational learning are:

1. Establishing a culture of continuous improvement: Cultivate a mindset of constant improvement at all levels of the organization, encouraging the active participation and commitment of all members.

2. Data Measurement and Analysis: Collect and analyze relevant data to identify areas of opportunity, set objectives and goals, and evaluate progress toward continuous improvement.

3. Training and skills development: Invest in the development of staff skills and knowledge, encouraging knowledge transfer and the adoption of best practices.

4. Learning from mistakes and successes: Promote an environment of openness and learning, where both mistakes and successes are analyzed constructively, extracting lessons and applying the knowledge acquired.

5. Collaboration and teamwork: Encourage interdisciplinary collaboration and teamwork, leveraging diverse perspectives and experiences to drive innovation and continuous improvement.

6. Committed leadership: Having a leadership committed to continuous improvement and organizational learning, providing support, resources, and creating a culture of excellence.

Continuous improvement and organizational learning are fundamental pillars for the sustainable success of any organization. By handling these concepts and applying them effectively, companies can stay competitive, adapt to changes, optimize their processes, improve the quality of their products and services, and foster an environment of constant growth and development for all their members.

Identification of Lessons Learned

- Thorough review of audit findings and non-conformities.
- Analysis of root causes and contributing factors.

- Evaluation of the effectiveness of the corrective and improvement actions implemented.

Compilation and Documentation of Lessons Learned

- Establishment of a structured process for the collection of lessons learned.
- Use of standardized tools and formats.
- Maintenance of a database or repository of lessons learned.

Analysis and Prioritization of Lessons Learned

- Evaluation of the potential impact of lessons learned.
- Prioritization of the most critical and relevant lessons.
- Identification of key areas for process and system improvement.

Development of Action Plans

- Setting specific objectives and goals
- Definition of necessary activities and tasks
- Allocation of responsibilities and resources

Implementation of Process and System Improvements

- Updating of procedures, work instructions and documentation in general.
- Training and effective communication to the staff involved.
- Follow-up and monitoring of the implementation of improvements.

Evaluating the Effectiveness of Improvements

- Measurement and analysis of key performance indicators.
- Conducting follow-up audits.
- Identification of areas that require additional adjustments.

Institutionalization of Lessons Learned

- Integration of lessons learned into organizational culture.
- Promotion of continuous learning and constant improvement.
    - Establishment of mechanisms for sharing and transferring knowledge.

Change Management

- Planning and control of changes resulting from improvements.
- Training and sensitization of staff on the changes.
- Monitoring and evaluation of the effectiveness of the implemented changes.

Continuous Process Review and Improvement

- Periodic evaluation of the process of using lessons learned.
- Identification of opportunities for improvement in the process.
- Fostering a culture of learning and continuous improvement in the organization.

It is paramount to leverage lessons learned from audits to improve an organization's processes and systems. Starting from the identification and collection of lessons learned to their analysis, prioritization and effective implementation. In addition, the importance of evaluating the effectiveness of improvements, institutionalizing lessons learned, and fostering a culture of learning and continuous improvement throughout the organization should be emphasized.

INTEGRATION OF CONTINUOUS IMPROVEMENT INTO ORGANIZATIONAL CULTURE

Continuous improvement is much more than a process or a methodology; It represents a comprehensive approach that drives constant evolution and excellence in organizations. The ability to integrate continuous improvement into the organizational culture is essential to promote an environment where innovation, efficiency and continuous learning are core values.

Organizational culture, defined by its shared values, behaviors, and beliefs, plays a critical role in determining how an organization addresses challenges and pursues excellence. When continuous improvement is ingrained in this culture, it becomes an inherent part of the way the organization operates and develops over time.

The successful integration of continuous improvement into organizational culture offers a number of significant benefits. From engaging employees in identifying opportunities for improvement to promoting the organization's agility and adaptability in the face of change and challenges, this integration translates into a sustainable competitive advantage.

In this context, we will explore practical strategies to promote continuous improvement as an integral part of organizational culture, including engaged leadership, employee engagement and training, promoting an environment open to experimentation and learning, as well as implementing effective feedback systems.

By integrating continuous improvement into organizational culture, organizations can enhance their ability to innovate, adapt, and continuously grow, thereby enabling them to consistently achieve and exceed their strategic goals and stay ahead of the curve in a dynamic and competitive business environment.

As a quality expert, here are some ideas for effectively integrating continuous improvement into organizational culture:

- **Engaged leadership:** Top management must actively lead and promote the culture of continuous improvement, setting clear goals, allocating adequate resources, and visibly and actively participating in improvement initiatives.

- **Training and skills development:** Invest in training and skills development for all levels of the organization, fostering the knowledge and competencies necessary for continuous improvement.

- **Effective communication:** Implement effective communication channels to disseminate the benefits, advancements, and achievements of continuous improvement, as well as to gather ideas and feedback from employees.

- **Recognition and rewards:** Establish a recognition and reward system that values and celebrates employees' contributions to continuous improvement, both individually and as a team.

- **Employee involvement:** Engage employees in identifying opportunities for improvement, solving problems, and implementing solutions, fostering their sense of belonging and commitment.

- **Improvement teams:** Form multidisciplinary teams dedicated to continuous improvement, working on identifying and resolving problems, as well as implementing improvements in processes and systems.

- **Tools and methodologies:** Provide training and resources for the use of continuous improvement tools and methodologies, such as Lean, Six Sigma, Kaizen, among others.

- ➢ **Data measurement and analysis:** Establish key performance indicators (KPIs) and collect data to measure and analyze the progress of continuous improvement initiatives.

- ➢ **Learning and knowledge transfer:** Encourage continuous learning and knowledge transfer through training sessions, workshops, mentorships, and communities of practice.

- ➢ **Integration into processes and systems:** Incorporate continuous improvement as an integral part of the organization's processes and systems, ensuring that it is a sustainable and culturally rooted practice.

- ➢ **Celebration of achievements:** Recognize and celebrate achievements and successes achieved in continuous improvement, reinforcing the culture and motivating employees to continue contributing.

- ➢ **Benchmarking and adoption of best practices:** Research and adopt the best practices of other leading organizations in continuous improvement, learning from their experiences and adapting them to the organizational context.

- ➢ **Sustainable cultural change:** Constantly working to create lasting cultural change, where continuous improvement is an integral part of the way the organization thinks and acts.

These ideas encompass different aspects, such as leadership, training, employee engagement, tools and methodologies, measurement, learning, and recognition. The effective implementation of these ideas will contribute to integrating continuous improvement into the organizational culture, turning the organization into an environment of learning, innovation, and constant improvement.

12 CASE STUDIES AND EXAMPLES

DETAILED ANALYSIS OF REAL CASES OF INTERNAL AUDITS AND TREATMENT OF NON-CONFORMITIES ACCORDING TO ISO19011:2018

Detailed analysis of real cases of internal audits and treatment of non-conformities according to ISO19011:2018, applied to a quality management system based on ISO 9001:2015.

Example 1:

Case 1: Lack of control of documents and records

During the internal audit, a non-conformity related to inadequate control of documents and records of the quality management system was identified. Specific findings were:

1. Failure to review and update outdated procedures:

- Found several outdated procedures that did not reflect the organization's current processes.

- Some procedures contained contradictory or inconsistent information.

2. Use of Outdated Versions of Forms and Records:

- It was observed that staff were using old versions of forms and records, which made it difficult to track and trace information.

- Some records were incomplete or contained erroneous information due to the use of outdated versions.

3. Messy files and records without a clear identification system:

- The physical and electronic archives of documents and records were in disarray and without a clear organizational structure.

- There was a lack of a consistent identification and coding system for documents and records.

- Some records were misplaced or could not be easily located.

Processing according to ISO 19011:2018:

a. Identification and description of the non-conformity: The non-conformity was documented in a clear and precise manner, detailing the findings and evidence found. Screenshots, photographs, and specific examples were included to support the findings.

b. Root Cause Analysis: The Ishikawa diagram was used to identify potential root causes, such as lack of staff training in document control, absence of adequate procedures for reviewing and updating documents, lack of resources and responsibilities allocated for file maintenance, and lack of awareness of the importance of document and record control.

c. Corrective Actions: The following corrective actions were established:

- Review and update of all obsolete procedures, with the participation of the corresponding process owners.

- Training of staff on document and record control, including the correct use of the latest versions.

- Implementation of a software-based document and records control system, with versioning, distribution and access functions.

- Reorganization and coding of all physical and electronic files according to a new identification system.

- Assignment of specific responsibilities for the maintenance and control of documents and records.

d. Follow-up and verification: Quarterly follow-up audits were scheduled to verify the effective implementation of corrective actions and evaluate their effectiveness in eliminating non-conformity. These audits included the review of a representative sample of documents, records, and archives for up-to-date status and adequate control.

e. Closure of non-conformity: After two cycles of follow-up audits, it was verified that corrective actions had been successfully implemented and that non-conformance had been effectively eliminated. A final report was presented to management, including evidence of the resolution of the non-conformity, and the resolution was formally closed.

Case 2: Failure to comply with product requirements

In another internal audit, a non-conformance related to non-compliance with specified requirements for a specific product was detected. The findings were:

1. Deviations in product characteristics from customer requirements:

- Discrepancies were found between the specifications agreed with the customer and the actual characteristics of the delivered product.

- Some key features of the product did not meet the requirements set by the customer.

2. Lack of validation of product requirements during design and development:

- There was no evidence of formal review and validation of product requirements during the design and development stages.

- Some customer requirements were not considered or were omitted during the design process.

3. Absence of records of verification and validation of requirements:

- No records were found demonstrating verification and validation of product requirements prior to release for production.

- There was a lack of evidence of the customer's formal approval of the design before proceeding with manufacturing.

Processing according to ISO 19011:2018:

a. Identification and description of nonconformity: The nonconformity was documented in detail, including specific findings and evidence found, such as inspection reports, incomplete design records, and customer complaints.

b. Root Cause Analysis: The cause-and-effect diagram (Ishikawa) was used to identify potential root causes, such as lack of effective communication with the customer during the requirements definition stage, deficiencies in the design and development process, lack of staff training in requirements validation, and absence of adequate controls for requirements verification and validation.

c. Corrective Actions: The following corrective actions were established:

- Comprehensive review of the design and development process, including the incorporation of mandatory requirements verification and validation stages.

- Implementation of a systematic communication process with customers during the definition of requirements, including the formal approval of requirements before proceeding with the design.

- Training of personnel involved in design and development on the importance of requirements validation and the techniques to do so effectively.

- Establishment of additional verification and validation controls, including review by independent experts prior to release to production.

d. Improvement Actions: In addition to corrective actions, the following improvement actions were implemented to avoid the recurrence of non-conformity in other products or projects.

- Review of the requirements of all existing products to identify potential non-compliances.

- Standardization of the design and development process across the organization, including the requirements verification and validation stages.

- Implementation of lessons learned and best practices identified during non-conformance treatment.

e. Follow-up and verification: Quarterly follow-up audits were conducted to verify the implementation and effectiveness of corrective and improvement actions. These audits included reviewing design records, evidencing communication with customers, and verifying the application of requirements validation controls on a sample of projects and products.

f. Closure of non-conformity: After three cycles of follow-up audits, it was verified that corrective and preventive actions had been effectively implemented and that the non-conformity had been resolved. A final report was presented to the management, including evidence of the resolution of the non-conformity and the implementation of the improvement actions, and the formal closure of the same was carried out.

Example 2:

Detailed analysis of real cases of internal audits and treatment of non-conformities according to ISO19011:2018, applied to an information security management system based on ISO 27001.

Case 1: Lack of access control to confidential information

During the internal audit, a non-conformity related to the lack of access control to confidential information of the organization was identified. Specific findings were:

1. Unauthorized Access to Shared Folders with Sensitive Information:

- Multiple employees were found to have access to shared folders containing sensitive customer information and sensitive financial data, without having the authorization or need to access such information.

- Some of these employees had accessed and downloaded files with no apparent justification.

2. Lack of restrictions on access to databases with customer information:

- Databases storing customers' personal information, such as names, addresses, and identification numbers, were found to not have adequate access restrictions.

- Any employee with access to the internal network could access and modify this data.

3. Absence of Critical Information Access Monitoring Logs:

- No audit trails were found that would allow monitoring and tracking access to critical and confidential information of the organization.

- There was no evidence that access to this information was tracked.

Processing according to ISO 19011:2018:

a. Identification and description of nonconformity: The nonconformity was documented clearly and accurately, detailing the findings and evidence found, such as screenshots, access logs, and employee testimonials. Specific details about the affected files and databases were included, as well as examples of unauthorized access.

b. Root Cause Analysis: The Ishikawa diagram was used to identify potential root causes, such as the lack of a clearly defined and communicated information access control policy, the absence of procedures for the assignment and revocation of access permissions, the lack of information security training of personnel, the absence of technical resources to implement effective access controls and the lack of adequate monitoring and follow-up by the organization.

c. Corrective Actions: The following corrective actions were established:

- Development and implementation of an information access control policy, including roles, responsibilities, access levels, and processes for assigning and revoking permissions.

- Thorough staff training in information security, awareness of the importance of protecting sensitive information, and access control processes.

- Implementation of technical access controls, such as role-based user authentication, folder and file-level access control, and access monitoring and logging.

- Assignment of specific responsibilities for the management and monitoring of access controls, including the periodic review of permissions and access.

- Thorough review of existing access permissions and revocation of unauthorized or unnecessary access.

d. Follow-up and verification: Quarterly follow-up audits were scheduled to verify the effective implementation of corrective actions and evaluate their effectiveness in eliminating non-conformity. These audits included reviewing access logs, verifying technical controls in place, testing for random access, and interviewing staff to assess their understanding of access control processes.

e. Closure of non-conformity: After two cycles of follow-up audits, it was verified that corrective actions had been successfully implemented and that non-conformance had been effectively eliminated. A final report was submitted to management, including evidence of the resolution of the non-conformity, such as screenshots of the controls in place, up-to-date access logs, and staff training records. The non-conformity was formally closed.

Case 2: Lack of Security Incident Management

In another internal audit, a non-conformity related to the lack of proper management of information security incidents was detected. The findings were:

1. Absence of a formal process for identifying, recording and dealing with security incidents:

- There was no established and documented procedure for the management of information security incidents.

- Staff did not have clear guidelines on how to identify, report, and handle security incidents.

2. Lack of records of security incidents occurred and actions taken:

- No records or documentation were found of security incidents that had occurred in the organization.

- There was no evidence that actions had been taken to investigate, contain, and resolve reported security incidents.

3. Lack of staff training in security incident management:

- Staff had not received formal training on information security incident management.

- There was a lack of knowledge and awareness about the importance of properly reporting and managing security incidents.

Processing according to ISO 19011:2018:

a. Identification and description of non-conformity: The non-conformity was documented in detail, including specific findings and evidence found, such as staff testimonials, incomplete or non-existent incident records, and examples of security incidents that were not properly handled.

b. Root Cause Analysis: The cause-and-effect diagram (Ishikawa) was used to identify potential root causes, such as lack of leadership and management commitment in incident management, absence of formal procedures and guidelines, lack of defined roles and responsibilities, lack of staff training, lack of awareness of the importance of incident management, and absence of tools and resources for recording and Incident tracking.

c. Corrective Actions: The following corrective actions were established:

- Development and implementation of a procedure for the management of security incidents, including stages of identification, recording, analysis, containment, eradication, recovery and lessons learned.

- Designation of a security incident response team, with clearly defined roles and responsibilities.

- Comprehensive training of staff in security incident management, including awareness of the importance of reporting and recording incidents, as well as the processes and procedures to be followed.

- Implementation of centralized tools and logs for tracking and documenting security incidents.

- Establishment of communication channels to report security incidents in an efficient and timely manner.

d. Improvement actions: In addition to corrective actions, the following improvement actions were implemented to prevent the recurrence of non-conformity:

- Regular review and update of the security incident management procedure, based on lessons learned and changes in the

These case studies demonstrate the importance of following a systematic and structured approach to the treatment of non-conformities, as set out in ISO 19011:2018.

13 CONCLUSION

As an advisor on ISO 19011:2018, I can say that it is critical for an organization's leadership to understand the crucial importance of audits and corrective actions as essential tools for business success. At the end, I want to share the most relevant tips, according to my experience.

1. Promote a culture of continuous improvement: Management should be the primary promoter of an organizational culture that values audits and corrective actions as opportunities for learning, improvement, and growth.
For example, you can:

- Constantly communicate the importance of continuous improvement and the benefits that come with it.
- Establish clear objectives and goals related to continuous improvement and management system performance.
- Foster an environment of openness and transparency, where staff comments and suggestions are encouraged.
- Recognize and celebrate achievements and successes through continuous improvement initiatives.

2. Allocate adequate resources: Audits and corrective actions require sufficient resources, including trained personnel, time, budget, and adequate tools. The address can:

- Conduct a thorough assessment of the resources needed to conduct audits and implement corrective actions.
- Allocate a specific budget for these activities, considering the costs of training, tools, external consulting (if necessary), among others.
- Ensure that the staff involved have the necessary time and availability to participate in these activities without compromising their day-to-day responsibilities.
- Invest in the acquisition and maintenance of tools and technologies that facilitate the management of audits and corrective actions.

3. Ensure the independence and impartiality of audits:
Audits should be conducted objectively and impartially to ensure the validity of their findings. The address can:

- Establish a clear and transparent process for the selection and assignment of auditors.
- Ensure that auditors do not have conflicts of interest or relationships that could compromise their objectivity.
- Foster a culture of ethics and integrity in the organization, where honesty and impartiality are valued in audit processes.
- Implement mechanisms to manage and address any potential conflicts of interest or threats to auditor independence.

4. Encourage the participation and commitment of all staff:
Management should promote the active participation of all staff in the audit and corrective action processes. This can be done as follows:

- Clearly and effectively communicate the objectives, benefits, and expectations related to these activities.
- Involve staff in the planning and execution of audits, as well as the implementation of corrective actions.
- Foster an environment of trust and openness, where staff feel comfortable expressing their concerns and suggestions.
- Provide adequate training and guidance to staff on their roles and responsibilities in these processes.

5. Prioritize risk-based corrective actions: Corrective actions should be prioritized based on their potential impact on the organization's risks and goals. The address can:

- Establish a structured process for evaluating and prioritizing nonconformities and corrective actions.
- Consider factors such as the severity of the non-conformity, its impact on critical processes, the associated risks, and the resources required.
- Involve the leaders of the affected areas in the prioritization of corrective actions.

- Allocate resources and assign responsibilities according to established priorities.

6. Monitor and review the effectiveness of corrective actions: Management should establish mechanisms to monitor and review the effectiveness of corrective actions implemented. For example:

   - Establish key performance indicators (KPIs) to measure the effectiveness of corrective actions.
   - Conduct follow-up audits to verify the implementation and effectiveness of corrective actions.
   - Periodically review reports and records related to corrective actions implemented.
   - Involve affected areas in evaluating the effectiveness of corrective actions.

7. Encourage learning and continuous improvement: Audits and corrective actions are valuable sources of organizational learning. The address can:

   - Establish mechanisms to capture and document lessons learned during audit and corrective action processes.
   - Promote the exchange and dissemination of these lessons through training sessions, workshops, or knowledge-sharing platforms.
   - Foster an environment of openness to change and innovation, where ideas and suggestions for staff improvement are encouraged.
   - Integrate identified best practices into the organization's processes, procedures, and policies.

8. Recognize and celebrate accomplishments: Management should recognize and celebrate accomplishments through audits and corrective actions. Example:

   - Establish a recognition and rewards program for teams or individuals who have contributed significantly to these efforts.

- Communicate and disseminate the successes and positive results obtained through these processes.
- Celebrate milestones and achievements, fostering a sense of pride and motivation in staff.
- Use these achievements as inspiring examples to drive a culture of continuous improvement across the organization.

9. Lead by example: Management must lead by example, demonstrating their commitment to audits and corrective actions through their active participation, making informed decisions, and allocating appropriate resources. Example:

- Personally participate in audit reviews and corrective action follow-up meetings.
- Make decisions based on audit findings and recommendations.
- Allocate adequate resources and budget to support these activities.
- Conspicuously and consistently communicate the importance of audits and corrective actions to the success of the organization.

An organization's leadership can take full advantage of the potential of audits and corrective actions as essential tools for business success. This allows the organization to identify areas for improvement, implement effective corrective actions, foster organizational learning, and ultimately improve its performance, efficiency, and competitiveness in the market in a sustainable manner.

14 BIBLIOGRAPHY

Standards & Guidelines:

1. ISO 19011:2018 - Guidelines for the audit of management systems.
2. ISO 9001:2015 - Quality Management Systems - Requirements.
3. ISO 27001:2013 - Information technology - Security techniques - Information security management systems - Requirements.
4. ISO 14001:2015 - Environmental Management Systems - Requirements with guidance for their use.
5. ISO 45001:2018 - Occupational health and safety management systems - Requirements with guidance for their use.

Books:

6. Russell, J. P. (2021). The ASQ Auditing Handbook: Principles, Implementation, and Use (4th Edition). Quality Press.
9. Kausek, J. (2019). Effective Auditing for Compliance: A Systematic Approach to Auditing Systems of Management. CRC Press.
10. Moeller, R. R., & Witt, H. N. (2020). Auditing and Assurance Services (16th Edition). Pearson.
11. Teixeira, S., & Sampaio, P. (2021). Integrated Management Systems: Leading Strategies and Solutions. Springer.

FINAL MESSAGE

Audits are a critical tool to promote continuous improvement, efficiency, and compliance in any organization. Through this book, we have explored in depth the principles, techniques, and best practices in internal auditing and non-conformance management.

We have learned that audits are not simply a compliance activity, but an invaluable opportunity to identify areas for improvement, evaluate the effectiveness of systems and processes, and promote a proactive approach to organizational excellence.

Throughout these pages, we have highlighted the importance of following a structured and systematic approach in planning, executing and monitoring audits. We have explored effective techniques for the collection and analysis of evidence, the communication of findings, and the implementation of corrective and preventive actions.

In addition, we have emphasized the importance of root cause analysis and adopting a risk-based approach to prioritize and address non-conformances effectively. We have highlighted the need to foster a culture of continuous improvement, where audit findings are perceived as opportunities for learning and organizational growth.

In the pages of this book, we've shared case studies, illustrative examples, and valuable tools to help you apply the concepts and principles of auditing in your own organizational environment. I hope this guide is a helpful resource and source of inspiration.

Remember that the success of audits and non-compliance management depends on the commitment and participation of all members of the organization. It is essential to have the support and leadership of senior management, as well as the collaboration and commitment of work teams at all levels.

Maintain an open mindset to learning and continuous improvement. Take advantage of the lessons learned from audits to optimize your processes, strengthen your management systems, and create sustainable value for your organization and all its stakeholders.

I am confident that this book has provided you with the tools and knowledge necessary to embark on a journey to excellence through effective audits and proactive non-conformance management. Remember that this is just the beginning of a path of continuous improvement that never ends.

Much success in your adventure towards operational excellence!

With optimism and support,

Rodrigo Palma Mena

www.ingramcontent.com/pod-product-compliance
Lightning Source LLC
Chambersburg PA
CBHW050109230526
45470CB00004B/1743